First Facts™

Why in the World?

Do Bed Bugs Bite?

A Book about Insects

by Pamela Dell

Consultant:
Gary A. Dunn, Director of Education
Young Entomologists' Society, Inc.
Minibeast Zooseum and Education Center
Lansing, Michigan

Capstone
press®
Mankato, Minnesota

First Facts is published by Capstone Press,
151 Good Counsel Drive, P.O. Box 669, Mankato, Minnesota 56002.
www.capstonepress.com

Library of Congress Cataloging-in-Publication Data
Dell, Pamela.
Do bed bugs bite? : a book about insects/by Pamela Dell.
p. cm.—(First facts. Why in the world?)
Summary: "A brief explanation of insects, including physical characteristics, life cycles, and
habitats"—Provided by publisher.
Includes bibliographical references and index.
ISBN-13: 978-0-7368-6785-6 (hardcover)
ISBN-10: 0-7368-6785-6 (hardcover)
1. Insects—Juvenile literature. I. Title. II. Series.
QL467.2.D45 2007
595.7—dc22   2006023562

Editorial Credits
Jennifer Besel, editor; Juliette Peters, set designer; Renée Doyle, book designer; Wanda Winch,
    photo researcher/photo editor

Photo Credits
Creatas, cover (pinching bug)
Digital Vision, 8–9
Dwight R. Kuhn, 16–17
Minden Pictures/Pete Oxford, 20; Stephen Dalton, 10
Nature Picture Library/Dietmar Nill, 21; Kim Taylor, 19
Peter Arnold/Luiz C. Marigo, 15
Shutterstock/Alexander M. Omelko, 4; Dainis Derics, 12; Jody Dingle, 11;
    Konrad Lewandowski, 5
University of Florida/IFAS Communication Services, cover (bed bugs)
Visuals Unlimited/Richard Walters, 6

1 2 3 4 5 6 12 11 10 09 08 07

Table of Contents

What Is an Insect?

If people did not rule the world, insects would. There are more insects than all the people and other animals on earth combined. But how can you tell insects from other creepy, crawly things? Read on to find out.

snout beetle

Scientific Inquiry

Asking questions and making observations like the ones in this book are how scientists begin their research. They follow a process known as scientific inquiry.

Ask a Question

Are ants insects?

Investigate

In spring or summer, look around your neighborhood for ants. When you find one, use a magnifying glass to study the ant's body. When you're done, read this book to learn what an insect is.

Explain

You noticed the ant's body had three sections. You saw six legs. You also saw two **antennas** sticking out of its head. You decide that ants are insects. Write it in your notebook and remember to keep asking questions.

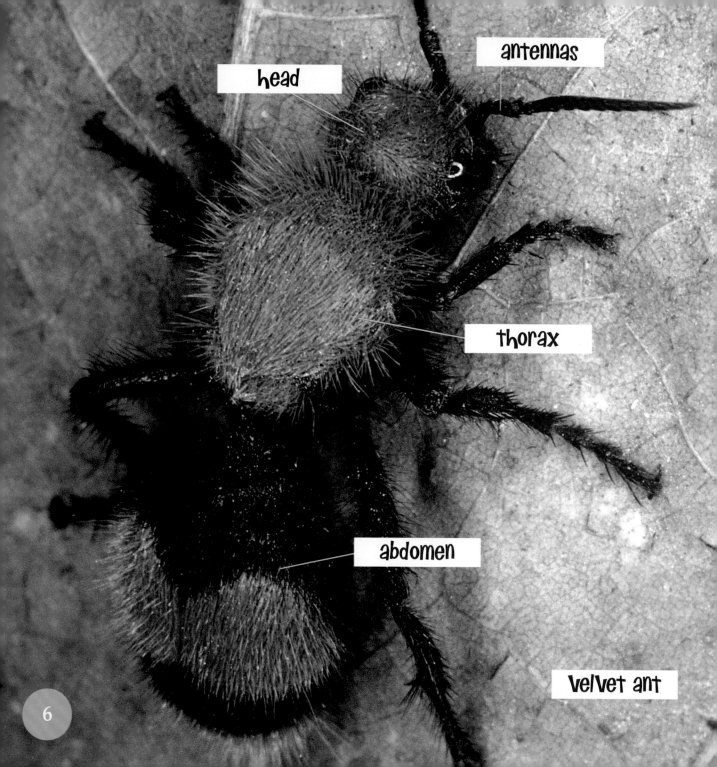

head

antennas

thorax

abdomen

velvet ant

6

All insects have six legs and two antennas. They also have three body parts. These parts are the head, the **thorax**, and the abdomen. Most adult insects also have wings.

Did you know that insects have a skeleton on the outside? It's called an **exoskeleton**. This hard outer shell protects an insect's soft organs.

? Did you know?
Not all buglike animals are insects. Spiders have eight legs and no wings or antennas. Spiders are arachnids.

Do Insects Have Homes?

Insects live almost everywhere. They make homes in plants, under rocks, or on the bodies of other animals.

Some termites build towering mounds of dirt to live in. These mounds keep the termites safe and at just the right temperature.

bed bug

Do Bed Bugs Bite?

Sooner or later, insects must eat. Bed bugs bite to get blood, their main food. Other insects, like mosquitoes and kissing bugs, bite for blood too.

Not all insects eat blood. Flies, honey bees, and butterflies feast on pollen or the nectar of flowers. Many beetles find hair, feathers, and skin simply delicious.

monarch butterfly

Do Insects Work?

They do! Insects spend their lives working. Their work is finding food, escaping enemies, **mating**, and producing young.

Insects that live in **colonies** work together at other jobs too. Inside their hive, honey bees make honey. Others are soldiers, protecting the hive. Some nurse the young.

? Did you know?

The job of some worker ants is to keep their nests clean. They take garbage out of the nest and put it in a dump outside.

How Do Insects Stay Safe?

Insects have many ways to fight off **predators**. The wood ant sprays acid from its rear that causes a burning pain in attackers.

Other insects stay safe by hiding. With **camouflage**, insects like the walking stick can't be spotted easily. If a predator can't find it, a predator can't eat it.

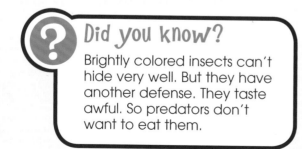

? Did you know?
Brightly colored insects can't hide very well. But they have another defense. They taste awful. So predators don't want to eat them.

Why Do Fireflies Light Up?

Nearly all insects must find mates to produce young. Those tiny lights that sparkle on summer nights are fireflies finding mates. Males flash their lights to get a female's attention. Then the female flashes back to say "Come on over."

How Long Do Insects Live?

Insects don't usually live very long. Some cockroaches live two years or more. But other insects have very short lives. Adult mayflies take to the air, mate, and die in only a day or two.

Insects die every day. But more hatch to take their place. Billions of insects are always among us.

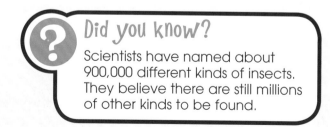

Did you know?

Scientists have named about 900,000 different kinds of insects. They believe there are still millions of other kinds to be found.

mayfly

People use their ears to hear, tongues to taste, and noses to smell. Not so for insects! Katydids and crickets hear using special organs on their knees. Flies, bees, and butterflies have supersensitive taste buds on the bottoms of their feet. Antennas come in handy for feeling things. But many bugs use their antennas for smelling too.

WHAT DO YOU THINK?

Insects have special jobs that help the earth in many ways. Many insects help keep our planet clean by eating dead matter. Other bugs protect plants by eating the insect pests that feed on them. What are some other ways that insects help the world?

GLOSSARY

antenna (an-TEN-uh)—a feeler on an insect's head

camouflage (KAM-uh-flahzh)—coloring or covering that makes animals, people, and objects look like their surroundings

colony (KOL-uh-nee)—a large group of insects that live together

exoskeleton (eks-oh-SKEL-uh-tuhn)—the hard outer shell of an insect; the exoskeleton covers and protects the insect.

mate (MATE)—to join together to produce young; a mate is also the male or female partner of a pair of animals.

predator (PRED-uh-tur)—an animal that hunts other animals for food

thorax (THOR-aks)—the middle section of an insect's body; wings and legs are attached to the thorax.

READ MORE

Aloian, Molly, and Bobbie Kalman. *Insect Life Cycles.* The World of Insects. New York: Crabtree, 2005.

Parker, Steve. *Ant Lions, Wasps, and Other Insects.* Animal Kingdom Classification. Minneapolis: Compass Point Books, 2006.

Solway, Andrew. *Deadly Insects.* Wild Predators! Chicago: Heinemann, 2005.

INTERNET SITES

FactHound offers a safe, fun way to find Internet sites related to this book. All of the sites on FactHound have been researched by our staff.

Here's how:

1. Visit *www.facthound.com*

2. Choose your grade level.

3. Type in this book ID **0736867856** for age-appropriate sites. You may also browse subjects by clicking on letters, or by clicking on pictures and words.

4. Click on the **Fetch It** button.

FactHound will fetch the best sites for you!

INDEX